Chapter 1

Getting to know your Fire HD

Starting the device

The first time you turn on your Fire HD you need to register your device. In order to turn your device on, you need to push the smaller button that is located at the top of the Fire HD and hold it for up to three seconds.

If you want to turn on the sleep mode, you will push the power button once and you will do the same in order to wake up the Fire HD. In order to turn the device off, you need to hold the power button for three seconds and a prompt will appear on the screen, allowing you to choose to restart, turn off or cancel.

Battery and controls

You can use the device for 12 hours when the battery of the Fire HD is fully charged, which is an upgrade from the 8 hours of the previous version.

On the top of the Fire HD, you will find a large button, on the opposite side of the power button. This is where you will adjust the volume. You will also find a 3.5mm headphone jack where you can plug in your earbuds or headphones.

Near the power button, you will find your charging port. This is where you will plug in your charger.

The device is a touch screen which means that you tap on option on the screen it is much like clicking the mouse on a computer. You only need to tap it once in order to select an icon, then wait for it to load. When you want to scroll, you will swipe your finger across the screen.

First time charging

When you first remove your fire HD from the box, the tablet should be charged enough for you to go through the set up as well as the registration.

You will also notice that your Fire came with USB cord which you can plug into your computer to charge it. However, it is important to know that this will take much longer than if you use a wall plug.

Set up your HD Fire and registration

1. After you see the welcome screen, you need to choose the

language you prefer. Then click continue.

2. It will be followed by a screen which lists all of the available wi-fi connections. Choose which wi-fi connection you will use and enter your password if needed.

3. Once you have a wi-fi connection, you will move on to the register your device screen. Here you will need to register your device with Amazon. You will need to enter your amazon account information and if you don't have an account, you will need to choose 'create an account.'

4. If you don't register your device, you will not be able to make purchases on Amazon.com or

access Kindle store. As you are registering your device, you need to enter your credit card information as your source of payment. This will allow you to make purchases through amazon and the Kindle Store.

5. You will also find that on the registration page you can click to read the terms of service. If you agree to them, click 'I agree' then you will click the register continue button.

6. Once this done, you will be taken to a new screen where you will need to select your time zone. It is important that you

select your correct time zone because if you do not, it could lead to issues with you connecting to your wi-fi network.

7. The next page will be a page on which you will confirm your account information. You need to select the person who will use the Fire HD, in this case, it will be your amazon name, you can have the option to add your child name here as well, so you can share the Fire HD with them and set educational goals and time limits. Then click continue

8. Once you have done all of this, you will have the option to link your social media accounts. This is completely optional, however, in order to link your accounts, you simply need to click on the social media button that you want to link, input your information and select "connect"

After that, just click continue to go to the next page, the next screen will ask you to enable your location.

When you finished all the setup, the welcome screen will appear, you can go through the on-screen tutorial now if you want to, or just exit to start using your Fire HD.

Storage and the Cloud Drive

There are two storage options when it comes to the Fire HD, 16GB, and 32GB. The truth is that either one of these is a fine option because when you purchase the FireHD, you will receive unlimited storage on the Amazon Cloud drive, that can be used for the items that you purchase through amazon.

It is important to know that this Cloud Drive cannot be used for media that you have copied from your computer. What it means that all of the music, audio books, and movies that you purchase from amazon, will be stored online instead of on your device.

I know that you are going in an area with no internet connection, you don't have to worry because you ca download the media that you want to take with you, use it while you have no connection and then delete it from your device. It will still be available through your cloud.

Learning to navigate

Let's begin with the navigation bar which is going to appear at the bottom of the screen no matter what app you are using, except when you are reading. The navigation bar will consist of three buttons which will look like a circle, a triangle and a square.

- The circle is going to take you home, or to the previous page that you selected.

- The triangle is going to take you back to the last screen that you were on.
- The square is going to take you to the task switcher which will allow you to see apps you have recently used as well as open or close them.

In order to check your wi-fi connection or show the quick settings option, swipe your finger down from the top of the screen and this screen below will appear.

It is important that you take the time to get to know you Fire HD before you begin using it so that if you do find yourself stuck you will know what to do and how you can return to your home screen menu.

Changing the Wallpaper

If you don't like the original wallpaper on your Fire HD, you can change it.

1. Go to settings, then click on display.
2. You need to select wallpaper and choose the available images that come with your Fire HD from there.

3. Or yo can choose your own images by clicking on pick image icon.

Chapter 2

Available settings

In order to access quick settings, you need to swipe down from the top of the screen. This will not only give you access to the quick settings, but it is also going to give you access to the more advanced settings on your Fire HD.

Or click the settings Icon on the home page.

Quick actions

The first thing that I want to go over with you are the quick actions that you can do simply by swiping down from the top of your screen.

- The first setting is the auto-rotate option. This setting will allow you to lock and unlock the auto-rotation o f the screen. If you want the screen to automatically rotate, you will turn this option on, on the other hand, if you do not want the screen to rotate, you can switch this to off.

- The brightness setting will allow you to adjust the brightness of your screen. The higher you set this setting the brighter your screen will be.
- The wireless setting os going to allow you to connect to a wi-fi network.
- Airplane mode setting is to take it off and on.
- Do not disturb setting is to mute all of your notification.
- The help settings are going to put you into contact with

an amazon customer service agent who will help you with any issue that you are facing.

- The setting option is going to take you into the more advanced settings.
- Blue shade setting is to turn the blue shade on or off.
- Camera setting allows you to turn the camera on.

More Advanced Settings

Once you choose the settings option, you are going to be able to go into the more advanced settings on your Fire HD.

1. The wireless option is going to enable you to connect to a wi-fi network and it is also going to enable you to turn the airplane mode on and off.
2. My account setting will allow you to connect to social networks as well as manage the accounts that are associated with the device.

3. Profiles and family library is the setting which will allow you to add profiles for the other members in the house. This is a great option if there are children that are going to be using the device.

4. "sync device" setting is going to allow you to sync your device and check for any updates.

5. The device options setting is going to allow you to update your device, back up the content that you have on your device, view the free storage space on the device, change

the date and the time that is displayed on your device and change the name of your device.

6. The power management setting is going to enable you to manage your battery usage.

7. The display and sound setting is going to enable you to adjust the brightness of the screen, adjust your volume

and it is going to allow you to adjust the mirroring settings.

8. The keyboard and language setting is going to enable you to change the language of the device as well as the keyboard and it is also going to enable you to choose the voice for the text to speech setting.

9. The parental controls are going to enable you to turn the parental controls both on and off, and you will be able to

manage the profiles for the children in your home.

All about Bluetooth pairing

You can pair your Fire HD via Bluetooth with other devices or accessories. In order to make Bluetooth pairing easier, you need to make sure before you get started that the device or accessory that you are trying to pair is within the range of your fire HD.

Begin by turning on Bluetooth for both of the devices and ensure that pairing has been enabled. In order to do this on your Fire HD, go on setting on your homepage and choose 'wireless'. From there, you will choose, "Bluetooth.'

Tap on to enable your Bluetooth. This will allow your Fire HD to search for other devices that are available. Under the list of available devices, choose the device that you want to pair your Fire HD with.

You will have to allow both of the devices to pair and then you are done.

Features

The mirroring feature (for only fire HD 10) is going to allow you to display what is in your Fire HD on your television or another compatible device. This can be done by pairing with the Fire HD or via an

HDMI dongle in your Fire HD is not able to find the device.

The first thing that you need to do when you are using the mirroring feature is to ensure that your device is discoverable to your Fire HD.

On your Fire HD, you should swipe down from the top of the screen and tap the settings option. From here, you will choose the display and sounds option and then the display mirroring option.

This will cause your Fire HD to look for any devices that are within the range that is compatible. Find the name of your device that you want to connect with and tap into it. The first time that you connect with a device, it could take up to 30 seconds.

In order to stop the mirroring, simply swipe down on your Fire HD screen and tap 'stop mirroring'.

Chapter 3

The internet and your device

Internet browser

Silk is the name of the internet browser that you will be using on your Fire HD. The Silk icon will be located on your homepage. In order to search for a website, simply type what you are searching for in the search bar and tap the search button.

On the options bar, you will also find a home icon, back and forward icons, a menu icon, and a favorites icon.

Silk works just like most of the browsers that you would use on your laptop or computer as it uses tabs, which will enable you to have more than one web page open at a time and switching between these tabs.

If you want to bookmark a site when you are using the silk browser, you will click on three dots icon, then tap on add bookmark. After you click this tab, a dialog box will appear where you can either cancel or bookmark the page. In order to delete a booked marked page, you will tap on the menu icon, then tap on the bookmark which will display a thumbnail of all the pages you have bookmarked.

If you are looking for a specific topic, within a page and do not want to read the entire page, click the three dots icon that is located in the options bar. A list of options will appear on the screen and you will choose, 'find in page.'

Tap on 'find in page' then type in what you are searching for. If what you are looking for is located within the page, you will be taken directly to it.

In order to view your browsing history, you will click the menu item that is located at the top of your screen on the silk browser. From here, you will tap the history button and then you can choose whether you want to see the history from today or the past seven days. If you see a web page that you want to visit, simply tap on it.

Setting up your Email

In order to set up your email account on your Fire HD, you will have to have an already established email account from an online provider such as AOL or Gmail.

Just tap on the email icon on your home page, then you can add your email account by input your email information.

Next, you will tap the save option and if you want to view your inbox, simply tap the view inbox option. You can set up as many email accounts as you want on your Fire HD.

How to use the Calendar

When you first open your calendar, you are going to see a completely empty calendar. At the top of the screen, you will see a tab which will allow you to choose how you want your calendar to be displayed... day, week,or month.

If you want to add an event to your calendar, you just need to click on (the green add sign), next the sync Amazon Cloud screen will pop up. It lets you know that Amazon Cloud calendar events will be synchronized across amazon devices and services. If you want to accept it then click accept. If you don't want to do that then click on setting, this will allow you to disable synchronization.

Now you're almost done, next screen is the screen that allows you to add an event to your calendar.

What about contacts?

The contacts app is pre-installed on the Fire Hd and you can access it by tapping on the app's icon which is found on your home screen. From there, you will simply tap on the contacts app.

If you already set up an email account, you can import all of your contacts from your associated email account. After you set up the email account and open up the contacts, a message will pop up on the screen asking you if you want to import your contacts.

If you want to add a new contact, simply tap the 'add contact' option and you will find fields for the name. address, phone number and email address for your new contact.

Tap on whatever field you wish to fill in and use the keyboard to fill in the information. If you want to add a , tap the icon on the side of the contact information. From here you can choose where you want to add the from. Tap the source that you want to get the from and when you are done, you will want to make sure that you click the save button.

Chapter 4

All about apps

How to download apps and game

There are free apps and games which you can download if they are available to your country. On your Fire HD, there will be an Amazon App Store app already installed.

Just simply select it, then you can choose the apps and games you want to download.

Once you have your choice, then go ahead and click the GET button, next click on download. Once it finished the download, the app icon will show up on your home screen. Now you can enjoy your app or game.

Learn how to uninstall apps and games

Now that you know how to download apps, it is also important for you to know how to uninstall apps and games. There are two different ways for you to uninstall apps and games for your Fire Hd.

The first way for you to uninstall an app is for you to hold your finger on the icon until a menu pops up which will give you the option to uninstall the app.

While that is the easiest way to uninstall an app, it does not always uninstall the app completely. If this happens, there is a second method which will allow you to ensure that app is completely uninstalled.

First, go to your setting, then tap on App and game. Next, you will need to go to manage all application, all your apps and games will appear here.

Now find app or game you want to remove, then tap on it and another screen will pop up. This is the screen which tells you all of the app's information and includes the UNINSTALL option. The last step is to click on uninstall, then this app or game will completely remove your Fire HD.

Features for everyone in the family

Not only is the Fire HD great for you, but it is great for the entire family, kids included! Because kids today enjoy spending time on devices, the Fire HD can make reading fun for them. It is great for kids because it is so easy to use and it offers parental controls to ensure that no matter what is downloaded, your child is safe and that they can only view content that is appropriate for their age.

Amazon offers an app called Free Time which focuses on children and educating them while they are still having fun. One of the great features of the app is that in order for your child to exit it, you have to enter a password which means that you will know exactly what they are doing while they are on the Fire HD.

However, the best thing about the app is that it allows you to set a time limit as to how long your child will be allowed to watch videos, play games and so on.

Of course, there are many different apps available for your Fire HD. It can be used to play games, watch videos, and of course read. It is finally the Kindle Fire that we have all been waiting for. The Kindle Fire is not just for Kindle books but is for everything that we do each day.

Chapter 5

Reading on your Fire

The Kindle was originally created for kindle book readers, but over the years, it has evolved into an amazing tablet that offers many features.

How to buy and read your books

Purchasing books and other media on your Fire HD is very simple. At the home screen, tap the book or newsstand icon, depending on which you want to purchase.

Then to go store tap on the shopping cart icon on the top right corner.

Scroll through and find what you want to purchase. Once you found what you want to purchase, tap on it. If you are purchasing a magazine or newspaper, you will tap the subscribe button.

If you are purchasing a book, you can preview the book to ensure that it is exactly what you are looking for. Tap'download sample' in order to see a preview. You will have to purchase the book in order to view the entire thing.

Once you decide that you want to purchase the book, tap the buy icon. After you purchase a book or magazine, it will automatically download to your Fire HD and it will also be stored in the cloud.

Accessing Free eBooks on Amazon

There are tons of free eBooks available on Amazon. The reason for this is because when an author first publishes a book, they are trying to get it in front of readers, therefore, they run promotions, giving the book away for free or at a reduced price. The authors able to run these promotions every three months, which means a lot of free books are available all of the time.

In order to access these free eBooks, all you have to do is type in the subject or genre that you want to read about, then sort the price from low to high. If you do not have Kindle unlimited, this is a great way for you to get plenty of eBooks and it is also a great way for you to decide if you want to purchase Kindle Unlimited.

If you do have kindle unlimited, all of the eBooks are free for you to read. This will allow you to access books written by all of your favorite authors for one low rate each month.

Reading Basics

While you are reading your books or media from the newsstand, you will swipe your screen from right to left to turn the page. If you have to stop reading, don't worry, the Fire HD will automatically bookmark the page that you are on then when you go back to reading you will begin on the last page that you were reading before you were interrupted.

You can also read and listen to your books at the same time as well as highlight the areas which are important to you or that you want to come back later. In order to highlight while using the Fire HD, you need to hold your finger down on the area that you want to highlight, then the highlighter icon will appear and you will tap it.

If you come across a word you don't know, then the Fire HD is going to automatically provide you with the definition of it, you can also translate it to other languages, and Fire HD also provide you with information from Wikipedia.

If you want to add a bookmark, tap on the center of the screen and you will see the icon bookmark on the right-hand corner. To add the bookmark, simply tap the icon.

In order to add a note, simply tap on the text that you want to highlight and then tap the 'note" option.

You will type in your note, and tap save. If you want to edit the note, tap the icon where the note appears, then tap edit. After you have made the changes that you want to make, simply tap save. In order to delete a note, tap the note icon then tap delete. A dialog box will appear and you will tap delete again.

Reading Kindle Book

1. Wordwise is a feature that you will use while you are reading your kindle books. You may notice on the products pages of some Kindle books that word wise is enabled. When you use

this feature, it will automatically detect words that are not commonly used and display hints on the page as to the meaning of the word.

This will make it easier for you to continue reading the books that you purchase without having to click on a word or open up the dictionary in another window. However, if you do want to see the full definition, you can click on the hints and the definition will be provided for you.

(For Fire HD 8): To turn word wise on and off, just simply tab on the center of your Kindle book page. And on the top right corner, you will see the three dots stacked icon, just click on it and choose an additional setting. It will bring you to the screens that allow you to turn it on or off.

(For Fire HD 10) In order to turn word wise on and off, just simply tab on the center of your Kindle book page. And on the top of your right-hand side you will see the three dots stacked icon, now just click on it and it will show Word Wise. Next, you need to click on it, and it will bring you to the screen that allows you to turn it on or off.

2. Word runner is a feature that is supposed to help you read faster because it displays one word at a time in the center of your screen. This feature will only work for books that have been written in English and it is not available with all titles.
In order to turn Word Runner on, while you are reading, you will tap the center of the page a

menu icon will appear that will look like three dots stacked on top of each other. Tap that icon and then choose Word Runner. From here, you will able to choose how many words per minute you want to be displayed with options ranging from 100 to 900 words a minute.

If you want to pause the Word Runner, simply hold your finger on the middle of the screen. While word runner is paused, you can swipe left or right to manually go through the words. When you are ready to start again, tap the center of the screen and the word runner will pick back up on the word displayed on the screen.

Chapter 6

How to Purchase and Listen to Audio

The Fire HD comes with amazon music already preinstalled which means that in order to access Amazon music, all you have to do is tap on the icon in order to download music, play music or purchase music. If you have an amazon prime membership, you can add your selected songs to a playlist, for free.

Adding Music

In order to add music from amazon to your Fire HD, you will simply select the music that you want to add from amazon music and tap the icon that looks like a plus sign.

If you push the play button while you are in amazon music store, the entire song is not going to be played, you will only hear a sample of the song. In order to listen to the entire song, you will need to add it to your music library.

Importing your music from another source

Many people have music on other devices such as their laptops or computers and they want to transfer that music to their Fire HD so that they do not have to repurchase it.if you want to do this, don't worry, it is quite simple to do.

It does not matter if you are trying to import music that has been purchased on iTunes or that has been ripped from your CDs, you will import it all the same way. However, it is important to know that you are only allowed to import 250 songs for free. After the first 250 songs, you will have to pay 24.99 per year to import up to 250,000 songs.

There are few different ways for you to import music from your laptop or computer to your Fire HD.

- The first way that I want to talk about is transferring your music by using a micro-USB cable. Begin by connecting both devices via the micro USB cable. Once the two are connected, you will want to unlock the screen on your Fire HD. Once you have unlocked the screen the computer or laptop is going to recognize that the Fire HD is connected to it.

Open your File Explorer and look under This PC/Computer and you will find the fire HD listed as Fire or Kindle. Highlight your music that you want to import and hold down the left mouse button, drag it to the Kindle or fire folder and put it in the music folder, dropping it there.

This will upload your music to the cloud and it will be

available through the music app on your Fire HD.

- The second way for you to import music from a computer or laptop. From the device that contains the music that you want to import to the fire had, go to Amazon music website. On the left of the screen, you will see the option, 'upload your music'. Click it. This will install the amazon music app onto your computer.

    Next, you will choose, "your library" which

is located at the top of the screen. Then look for "view the music on your computer" and click there. Now you will be able to right click on any of the music that is on your computer and upload it. Again, this will be uploaded to the cloud and will be available through the music app on your Fire HD.

How to listen to your Music

Now that you know how to get music onto your Fire HD, it is important for you to know how actually listen to that music. The first thing that you will want to do is to find the Music icon on the home screen and tap it.

Next, you will locate the music in your library that you want to listen to. Tap the song that you want to hear. If you tap the first song in a playlist, the Fire HD will play the entire playlist for you.

At the bottom of the screen, you will see the controls which you can use. These are previous, pause/play, shuffle and restart and volume.

How to listen to audiobooks

One of the great things about Kindle books is getting to listen to them, these are known as Audiobooks. From the audiobooks library that is found on your Fire HD, you will be able to purchase audiobooks, browse audiobooks and listen to audiobooks.

From the home screen, you will look for the audiobooks icon. In order to browse or purchase audiobooks, you will click on the shop option and find the book that you want to purchase.

If you tap the title you will be able to listen to a sample of the book, however, you must purchase the book in order to listen to it in its entirely.

If you want to listen to one of your audiobooks, swipe to the left side of the screen and tap on the audiobooks option. Here, your audiobooks will be listed. Simply tap on the audiobook that you want to listen to and it will be downloaded.

Once the audiobook is downloaded, you will be able to listen to the book, whether or not the Fire HD is connected to the internet.

How to manage your Audio

All of your audio files are going to be stored on the Amazon cloud. However, when you download, for example, an audiobook to your Fire HD, you will be able to listen to it without Wi-Fi access. After you are done listening to the audio, you can delete it from your Fire HD without deleting it from your cloud. This means that you will never have to repurchase it.

While I do want to focus on managing your audio, I also want to focus on managing all of your storage. All of the digital purchases that you make from Amazon are going to be stored on the Amazon cloud which will allow you to download them over and over again.

The reason for this is so that all of the storage on your device is not used up by the purchases that you make. One way to ensure that you are not wasting your storage is to delete the digital purchases from your device yu are done with them. Since these purchases are stored in the cloud, if you want to access them later, they will be there without taking up storage on your Fire HD.

If you want to remove digital purchases from your Fire HD and store them in the cloud all you have to do is go to settings>storage.

In this section, you will be able to archive items that have not been used recently to free up as mush space as you need in your device.

In free Time app if you download a book then it will be stored in your storage space on your Fire HD. If you want to remove and free up some space from your Free Time app you need to simply tap on the item that you want to remove and hold your finger on the item and after some time you will see a dialog box will pop up . in this way you will be able to click on 'Uninstall'. Keep in mind even if you remove the item from your device still it will be stored in the cloud.

Playing Music On Other Devices

You can also play music from the cloud to your devices. Simply log in to your Amazon account from your device and choose the music you want to listen or download or listen to.

How to Watch Videos

You can also watch videos on your Fire HD, with Fire HD device you will be able to purchase or rent then and you can also stream them to your TV.

Watching YouTube Videos

You can watch YouTube videos on your Fire HD, simply tap on the skill browser on your home screen and go to YouTube website. In this way, you can watch any video which you want to watch.

How To Open and Play Videos?

Simply go to the home screen in the Videos icon simply tap it. Now click on the menu icon which will be located at the top left-hand corner. This will bring up the Video Library.

This will show you all the videos you have rented, that has not expired, as well as all of the videos that you have purchased. In this way, you will be given the option to watch or download the desired video. You can watch or resume video you want to watch.

How can you buy or rent videos?

In order to access the videos, you have to purchase them or rent them. In order to do this, you will have to have a 1-click payment method set up. However, prime tv and videos can be watched at no additional charge. If you are a prime member, you can access tons of videos and television shows completely free.

If you renting a video, you will pay a specified amount which will allow you access to the video for a specific amount of time.

From the home screen, you will tap the Videos icon, then search for the movie or televisions show that you want to watch. If you know what you want to watch, simply tap the magnifying glass and type the name into the search bar.

If you are browsing in the store, simply swipe from the left side of your screen to look for movies or televisions shows. Once you have found what you want to watch, you will tap the rent or buy button.

After you purchase or rent your video, you can either stream it and watch it right then or you can download it and watch it later with no Wi-Fi necessary.

How to transfer your own videos to your Fire HD

Transferring your own videos to your Fire HD is not as complicated as it may sound. The first thing you will need to do is connect your computer to your Fire HD via a Micro- USB cable. Unlock the screen on your Fire HD, allowing your computer to recognize that the two are connected.

Open the file explorer on your computer and find where Kindle or Fire is listed. It is important for you to know where this is because this is where you will be placing the videos.

Next, highlight the videos you want to transfer and drag them to the

Kindle or Fire under movies folder. It will take your computer a few seconds depending on how many files you want to transfer. After you transfer these to the Fire HD, they will be located in the cloud, however, you will be able to download them to your device just as you would any other media.

Watch your Movies on your TV

The first way you can play the movies on your television is via an HDMI cable. To begin, you will need an HDMI cable with a standard USB connector at one end and a micro-

HDMI connector at the other. If you want to keep your Fire HD near you, enabling you to use the Fire HD as a remote, you will need a long enough cord to reach from your television to where you will be sitting.

It is important for you to know that this is only going to work if you have a television with an HDMI port. It is also important for you to understand that this is only going to work if you are using the Kindle Fire HD. Earlier versions of the Kindle and the HDX are not going to be able to be used with this method.

Begin by plugging the standard HDMI end of your HDMI cord into an HDMI port in your television. Plug the micro-HDMI end of the cord into your Fire HD.

It is important for you to make sure that your Fire HD is turned on before you switch to your HDMI input because if not, the television is not going to recognize the connection.

Even if you have a non-HDMI television, you can still watch the videos on your Fire HD on the television.

The first thing that you are going to need is to grab supplies. You will need an HDMI cable and a converter box.

It is important that you purchase a converter box that is labeled as HDMI to AV composite video + audio converter. It is also important that you purchase an HDMI cord that has standard HDMI connector at one end and the micro –HDMI on the other.

Begin by plugging the HDMI cord into your Fire HD and the converter box. Then plug the converter box into your television.

Alternatively, if you have a smart television, you can use the mirroring feature to play the videos on your television. In order to do this, you have to first ensure that your TV is discoverable to your Fire HD.

Chapter 8

All about your camera and documents

The fire HD comes with a front-facing camera that can be used for video calls and also can be used to take pictures and record personal videos.

It's important to know that all of the s that you take with the Fire HD are going to be uploaded to the cloud automatically.

In order to stop this from happening, you will need to tap the menu icon, while you are in the s library in then to settings. Here you'll see that option to turn automatic uploads on or off.

Transferring s

You can transfer your s from your Fire HD to your computer by connecting via a USB cable. Now, to transfer the s from your computer to your Fire HD select the s that you want to transfer. If you want to transfer more than one , you can hold down the CTRL button as you select the s. Drag the s to the Fire or Kindle folder and drop them in the s section. It will take your computer a few seconds to a few minutes to transfer the files depending on how many you are transferring.

In order to transfer pictures to your computer from your Fire HD, you need to click on the Fire or Kindle folder that is located under the computer folder. Select the s that you want to transfer using the same method mentioned above and drop them into whatever folder you want them to be located in on your computer.

Learning to Work with Documents

One app that you will notice on your home screen is the Docs library. This is where all your documents are going to be stored.

In order to get a document to your Fire HD, you will have to sideload it from your PC by connecting your fire HD to your Pc through USB cord. The just drag and drop the documents you want to transfer to the folder named "documents" which is located in the Fire or Kindle folder under the computer folder. And you can find the document you just transfer in the Docs Library under the local storage.

When you email a document to your Amazon email, it will automatically be uploaded to the cloud. This means if you have a document that you do not want automatically uploaded to the cloud, you need to turn this setting off before sending it. While you will be able to view your documents after they have been transferred to your Fire HD, you will not be able to edit them. The only way that you will be able to edit these documents is if you download an app such as OfficeSuite Pro.

Chapter 9

Troubleshooting

- **Keeps turning off**

   One common problem that many people face with the Fire HD is that it will turn off while it is used.

   1. The first thing that you want to do in order to fix this is to make sure that the battery is fully charged. **If the battery is charged**, then you will want to manually turn the Fire HD off by holding the power button for about 20

seconds. After the Fire HD is off, leave it off for about one minute then hold the power button down for 40 seconds turning it back on again.

2. **If the device is not turning off** but the screen is going dark, you will need to adjust your settings. Go to settings>Display>display sleep. After you have reached this location, you will want to increase the amount of time that it

takes for your screen to time out.

3. **Check to see if the Fire HD feels hot.** Sometimes, if you have a case on or have been using it for a long time, it can get overheated. Take the case off and turn it off. Set the Fire HD to the side and let it sit for a while until it cools off. Once the Fire HD had cooled, turn it back on and see if the problem is fixed.

4. **Check to make sure that your charger** and your

cable are working properly. Try changing the charger as well as the cord and see if this fixes the problem. I stated earlier that you have to be very careful when you are plugging in and unplugging your charger because you can ruin the cord.

If this happens, the fire HD will not work properly and even if you are using it while it is charging, could turn off in the middle of use.

5. **The last troubleshooting option** that you have is to backup all your data and then do a factory reset. In order to perform a factory reset, go to settings>device option>reset to factory defaults and then choose reset.

   If none of this works, you will need to call the Amazon customer service.

- **Keyboard types erratically**

Sometimes it has been reported that when the type Fire HD, skips the letters that are typed. And instead, it types random characters, deletes words, or skip pages while reading.

1. If you are having this problem then the first thing you want to do is to clean the screen. But keep in mind that when you clean the screen you must use a microfiber cloth. Also make sure that if you

are using a case on your Fire HD, it is fitted properly. Also, make sure that you are using a screen protector and there are no bubbles underneath.Try turning off the Fire HD. Hold the power button down for 20 seconds and then allow the Fire HD to keep turned off for 1 minute and then turn it on again. Also, make sure that the battery is charged if it does not work.If the above tips do not work, then back up all your data and do the factory reset. In order to do the factory reset simply go to settings>Device Options,>reset to factory defaults and then choose reset. If still, the problem persists then you need to call Amazon customer care.

www.ingramcontent.com/pod-product-compliance
Lightning Source LLC
Chambersburg PA
CBHW061148180526
45170CB00002B/676